The Cause of Menopause
&
Mercury Is Not a Planet

Toby L. Murray RMT

authorHOUSE®

AuthorHouse™
1663 Liberty Drive
Bloomington, IN 47403
www.authorhouse.com
Phone: 1 (800) 839-8640

© 2015 Toby L. Murray RMT. All rights reserved.

No part of this book may be reproduced, stored in a retrieval system, or transmitted by any means without the written permission of the author.

Published by AuthorHouse 07/22/2015

ISBN: 978-1-4969-4828-1 (sc)
ISBN: 978-1-4969-4827-4 (e)

Library of Congress Control Number: 2014918762

Print information available on the last page.

Any people depicted in stock imagery provided by Thinkstock are models, and such images are being used for illustrative purposes only.
Certain stock imagery © Thinkstock.

This book is printed on acid-free paper.

Because of the dynamic nature of the Internet, any web addresses or links contained in this book may have changed since publication and may no longer be valid. The views expressed in this work are solely those of the author and do not necessarily reflect the views of the publisher, and the publisher hereby disclaims any responsibility for them.

Contents

Preface vii
The Cause of Menopause.. 1
Alzheimer's, Parkinson's, Aging in- general,
and the Lunar-recession.. 8
Space travel and the Physiological Effects
of the Lunar-recession 16
Mercury is Not a Planet 19
Venus and Earth as 'Twin Sister' Planets 25
The Atomic Planets 35
Mars, Ceres, and the Asteroid belt................................ 37
Jupiter 38
Saturn 40
Uranus and Neptune.. 41
The Kuiper belt and Pluto 43
Bibliography.. 47
About the Author.. 55

Preface

Primary Respiratory Motion also known as the "Breath of Life" was first detected in the early 1900s by the American Osteopath Dr. William Sutherland. This motion arises out of the physiological system known as the *Primary Respiratory Mechanism* which includes the cerebral spinal fluid, the brain, spinal-cord, meninges, and the subtle movement which occurs at the *suture joints* between the bones of the human *cranium*. The Breath of Life expresses itself as three palpable wavelike motions called the *mid-tide, the long-tide*, and *the craniosacral-rhythm* which occur within every person's *central nervous system* with frequencies ranging from several seconds to over a minute. Each of these motions is palpable superficially on all living humans, between their cranium and sacrum as cerebral spinal fluid ebbs and flows within the central nervous system similar to the tides that occur in the ocean thus *osteopaths* and *craniosacral therapists* refer to the group of motions as "the three tides". Practitioners are able to determine whether each tide is slower, faster, weaker, or stronger than is optimal to diagnose and treat a variety of ailments such as allergies, migraines, and seizures (Kern, 2006, pg. 13), (Upledger, 1997, pgs. 209-211).

There is debate over what generates the palpable tides within cerebral spinal fluid. Traditional Chinese Medicine has asserted that human cerebral spinal fluid (which has a viscosity similar to egg-white) is controlled by 'kidney energy' and that each person's kidneys are governed by the Moon (Kern, 2005, pg. 43). Dr. Sutherland believed that the rhythmic fluctuations were governed by the "same great intelligence that governs the ocean's tides and the motion of the planets" (Ibid). In *Craniosacral Rhythm* therapist Daniel Agustoni suggested a relationship to 'celestial bodies' when he proposed that cerebral spinal fluid "is the body's inner-ocean and one has to consider in what proportions Earth, the Sun, and the Moon influence these rhythms" (2008, pg. 15).

The first portion of the following work combines Primary Respiratory Motion and the Biological Tides Theory with *the Inverse Square Law* to support the hypothesis that the Moon's gravitational pull upon cerebral spinal fluid is the principal source of Primary Respiratory Motion and as the Moon moves toward and away from Earth, cerebral spinal fluid is affected by its gravity enough to cause the cranium to expand and contract thus generating a Monthly Cranial Respiration Cycle which regulates the secretion of the brain's *neuro-endocrine hormones.*

The Cause of Menopause

Isaac Newton's *Universal Law of Gravitation* known as the *Inverse Square Law* explains how the Moon's gravitational pull is responsible for Earth's ocean tides as it physically lifts water that is directly underneath and closer to the Moon, higher than it lifts water which is located farther away. Hill & Playfair pointed-out in *Cycles of Heaven* that the human body is comprised of seventy-percent water and that all bodies of water respond to even the slightest fluctuations in the force of gravity (1978, pg. 31). In *How the Moon Affects You* psychiatrist Dr. Arnold Lieber presents the *Biological Tides Theory* suggesting that there are human physiological processes which ebb and flow in direct response to the Moon's gravitational tides. Lieber explained that there are biological high and low tides which are governed by the Moon throughout all of nature and that certain biorhythms within human physiology, particularly those pertaining to reproduction and emotional drives such as aggression, have a tendency to ebb and flow around the cycle of the 'new' and 'full' moon (1996, pg. 10). Emotional drives and reproduction are regulated by the neuro-endocrine hormones from the *pituitary gland*, the body's master neuro-endocrine hormone control-center which is located at the anatomical center of the brain and cranium. Osteopath Dr. Michael Kern explained in *Wisdom in the Body* that the rhythmic cycles of Primary Respiratory Motion gently rock the pituitary gland helping to milk the gland thus playing a role in regulating the secretion of the neuro-endocrine hormones (2005, pg. 71).

When Primary Respiratory Motion and the secretion of these hormones are considered together with the Biological Tides Theory, the Inverse Square Law, and the Moon's movements it is conceivable that each person's Monthly Cranial Respiration Cycle commences some time after the Moon begins to move closer to Earth and the steady increase in the pull from its gravity initiates *cranial inhalation* or an increase in the presence of cerebral spinal fluid in the central nervous system causing the expansion of an the cranium. *Cranial exhalation* commences after the Moon begins to move away from Earth and the overall strength of its gravitational pull starts to steadily decrease causing an individual's cranium to contract, intra-cranial pressure to increase, and cerebral spinal fluid to be propelled toward their pituitary gland. As the nine sets of paired cranial bones internally and externally rotate throughout primary respiration, spiraling currents of cerebral spinal fluid are directed toward either the anterior, intermediate, or the posterior pituitary lobe. Alterations in the flow of the cerebral spinal fluid and hydraulic-pressure directly upon the different pituitary lobes, provide the timing cues which are necessary for the secretion of the nine stimulatory and inhibitory neuro-endocrine hormones which regulate every person's growth, hydration, temperature, metabolism, puberty, pigmentation, emotional-drives, and the most prominent monthly hormonal cycle - menstruation.

The hormonal sequence of menstruation is regulated by the Moon's gravity via a Monthly Cranial Respiration Cycle similarly to the following brief outline. The average North

American girl experiences *menarche* - her first menstruation - at approximately twelve-years of age (U.S. Health, 2011). By the time she is between eighteen and nineteen years old she will stop growing at the average height for North American women (1.62 meters) and her pituitary gland will be situated along her brain's mid-line approximately six-centimeters below the top of her skull in the middle of the base of her brain which would be measured to be about 1.56 meters above the ground when she is standing. The average woman's menstrual flow may commence at either of the two times of the month that a statistically significant number of women's periods begin, at 'new' or 'full' moon (Lieber, 1996, pg. 70). When the Moon is 'new' on the first day of each of its 27.5-day *anomalistic* months, it is at its farthest point from Earth (apogee) and its gravitational attraction is at its weakest. As the Moon moves closer the steady increase in the pull of its gravity initiates the inhalation of cerebral spinal fluid into a woman's central nervous system, the expansion of her cranium, and the external rotation of her paired cranial bones. This directs the flow of cerebral spinal fluid and hydraulic-pressure toward her posterior pituitary lobe permitting the secretion of all six of the anterior lobe hormones including the *luteinizing* and *follicle stimulating hormones.* In daily doses these two hormones cross from a woman's central nervous system into her body where they work to induce a surge of the two ovarian hormones *estrogen* and *progesterone* until their concentrations spike near the end of the *follicular* or growth phase, and the same time of the month that the Moon reaches the mid-point of its orbit and its nearest approach to Earth (perigee) during each of its elliptical trips around Earth. At perigee the Moon is 'full' and more than fifty-thousand kilometers closer to Earth while its gravitational pull upon the average woman is over a *billion times* greater than it is when the Moon is at apogee. Immediately after its perigee, the Moon begins to move away and a woman's cranium starts to contract which creates an *isentropic compression* meaning that as her intra-cranial pressure increases there is an increase in her basal body temperature shortly before she ovulates and the exhalation of cerebral spinal fluid into her body begins, only moments prior to day-fourteen of her twenty-eight day reproductive cycle. At the same time, her brain's ventricles become taller and narrower and her paired cranial-bones internally rotate to redirect the flow of cerebral spinal fluid toward the anterior pituitary lobe, permitting the secretion of the only two posterior lobe hormones *oxytocin* and *anti-diuretic hormone* which combine with estrogen and progesterone to initiate the kidneys to begin retaining water in preparation for a pregnancy or menstruation and the next Monthly Cranial Respiration Cycle.

All North American women do not menstruate in unison due to the fact that each woman's personal relationship with the Moon's gravitational pull is 'relative' meaning that it is based on her height, body mass, and the exact physical distance that the Moon is from her pituitary gland and that average healthy women of different heights and sizes will have menstrual cycles with similar duration or *period* (28 days on average) but 'phases' that are not in sync with each other due to their different starting points. For instance, a tall large woman's cycle may begin on a particular day when the Moon is a given distance away, whereas a shorter and petite woman might not begin to menstruate until two or three days later, after the Moon has moved several thousands of kilometers closer. Though, as occurs naturally with all cycles, the phases of their menstruation will converge and diverge throughout the calendar year. Dr. Lieber explained that while assuming a lunar-timing influence on the female reproductive cycle, it must be understood that all women do not menstruate at the same time due to the fact that there is also a relative *solar* or daily component and that the different dates and times of each woman's birth are determining factors (Lieber, 1996, pg. 71). The daily component arises from the fact that the neuro-endocrine system includes the *pineal gland* another smaller midline gland that is situated only a few centimeters behind and above the pituitary gland. This gland is the source of daily secretions of the neuro-endocrine hormone *melatonin* which regulates the inflection of the sleep/wake and other daily body cycles known as the *circadian rhythms*. Melatonin secretion from the pineal gland is likely associated with a *Daily Cranial Respiration Cycle* that 'transmutes' or increases in frequency to become 'the three tides' permitting the pineal and pituitary glands to work in tandem and deliver the precise amount of each of the neuro-endocrine hormones that an individual needs on any given day of the month throughout life.

For the average North American woman the hormonal sequence of menstruation repeats in alignment to the lunar-cycle until approximately age fifty-one, the average age for the onset of *menopause* - the gradual decrease in the amount of reproductive hormones secreted from the pituitary gland leading to a complete absence of menstrual flow. Physiological changes such as dehydration, hot-flashes, and sudden mood swings precede full-menopause while at approximately the same age; the average North American man experiences the physiological equivalent, known as *andropause.* Classical Newtonian Mechanics dictate that all objects which are moving under the influence of a centrally directed force varying with the square of the distance must orbit in an elliptical path that *does not vary with time.* Given that scientists know that this type of unvarying relationship is shared between Earth and the Moon, the average person might expect to receive a consistent amount of neuro-endocrine hormones from their pituitary for millennia. Why then is it that menopause and andropause set in after only four decades? Does the functioning of the Monthly Cranial Respiration Cycle diminish? If so, why has human evolutionary physiology ensured that one's neuro-endocrine hormonal secretion is so finely regulated by the most reliable source known (gravity) only to have this vital physiological process cease functioning after such a relatively short-span of time? What is it about the average woman's body that changes drastically enough between the ages of eighteen and fifty-one (whether she ever gives birth or not) that it causes her reproductive capability to come to an end? Could it be the result of a change in her body mass? Over time, some women's body mass increases and other women's body mass decreases, but not consistently enough to explain the occurrence of menopause.

It is difficult to imagine that there is an aspect about the Moon that alters enough during the reproductive lifespan of the average North American woman to cause the functioning of her Monthly Cranial Respiration Cycle, to wane. Scientists know that the Moon is a 'dead world' and that nothing has changed about its mass or the composition of its core and lithosphere for over four-billion years. However, in *Orbiting the Sun* the astronomer Dr. Fred Whipple revealed that with the aid of light-reflectors left on the Moon's surface by the Apollo astronauts, lunar scientists were able to determine that the Moon's average distance from Earth is steadily increasing in a process known as the *Lunar-recession.* Dr. Whipple explained that the Lunar-recession occurs as a result of the Newtonian Mechanical law known as the *Conservation of Angular Momentum - as Earth's axial-rotation slows down the Moon's orbital-speed increases.* He specified that it has always been occurring and that it takes place at a rate of nearly four-centimeters each year (1981, pg. 17). It can be calculated that from the time of menarche to menopause (thirty-nine years on average) the Moon recedes away from Earth by 4cm x 39yrs, or roughly 1.56 meters (an amount that is similar to the measurement from the average North American woman's pituitary-gland to the ground). When the increase in the Moon's distance from Earth is factored into the relationship that the average woman shares with the Moon, it is evident that the Moon's gravitational pull upon her at age fifty-one has decreased to *half* of what it was when she reached adulthood and stopped increasing in height between the ages of eighteen and nineteen years old. Thus, the tidal-volume of cerebral spinal fluid within each of her Monthly Cranial Respiration Cycle along with the capacity of her cranial suture joints to move and permit her cranium to expand and contract and induce the secretion reproductive hormones from her pituitary gland will have diminished proportionately, causing the onset of menopause.

Alzheimer's, Parkinson's, Aging in-general, and the Lunar-recession.

It is while a fetus is in the anionic-fluid filled environment of the mother's womb that the Moon's gravity establishes the regulation of its Monthly Cranial Respiration Cycle and secretions of its neuro-endocrine hormones. In this way, both fetus and mother can physiologically respond to the Moon's movements independently of each other, while at the same time contribute genetic information to determining the length of the gestation period – from the point of conception to the time when the fetus` mass together with the amniotic fluid and the Moon`s gravity *perturb* the placenta enough to initiate labor. Dr. Lieber noted studies which found that women generally give birth at the time of the full-moon more than at any other time of the month (Lieber, 1996, pg. 71). Once a child is born the Moon's gravity begins to act directly upon its cerebral spinal fluid causing the developing cranial bones and the *fontanel ligaments* to expand and contract and grow in direct response to the regular increase and decrease in the strength of the Moon's gravitational pull arising from its elliptical movement in relation to Earth. During their developmental stage a child's cranial bones do not possess an *inherent growth potential* meaning that new bone-tissue is produced at the edges and stimulated to grow only in response to external stimuli, such as the cranial expansion (Opperman, 2000, pg. 2) caused by the inhalation of cerebral spinal fluid into the central nervous system resulting from an increase in the pull of the Moon`s gravity. Thus, the human cranium can be considered to be fully formed only after it has been motivated to expand and contract in direct response to the full progression of the Moon's repeating orbital pattern within its 18.6 year *Metonic* cycle.

After approximately two complete Metonic cycles and a period of near perfect cellular renewal in humans, cellular regeneration begins to take place slower than cellular deterioration as the Lunar-recession and the accompanying atrophy of cranial suture motion cause a marked reduction in the secretion of the neuro-endocrine hormones that have not yet been mentioned [*adrenocorticotropic hormone, growth hormone, thyroid stimulating hormone, and melanocyte stimulating hormone*]. This means that the Lunar-recession and the resulting diminishment of the Monthly Cranial Respiration Cycle are not only the cause of menopause but of the physiological losses associated with *senescence* or 'aging in-general' such as grey and thinning hair, slower metabolism, wrinkled-skin, a decrease in muscle-tone, and memory, vision and hearing loss; with the example of the average fifty-one year old North American woman as compelling evidence. After the passing of another thirty-nine years the Moon will have receded an additional 1.56 meters away from her and, factoring in the inverse square law, the overall strength of its gravitational pull upon her at the age of ninety will have diminished by *half* compared to what it was at the time that she entered into menopause at age fifty-one, leading to further atrophy of the motion of her cranial suture joints and a proportionate reduction in her cranium's ability to expand and contract and induce the secretion of neuro-endocrine hormones from her pituitary gland causing her bodily tissues to experience senescence and inevitably die of advanced age.

Dr. Sutherland believed that "living tissue is distinguished from tissue which is not living based on its property to express the motion of the Breath of Life and that without it there is no other bodily motion" (Kern, 2005, pg. 4). Thus, aging in-general and death due to advanced-age are merely consequences of the fact that we humans stop increasing in height between the ages of eighteen and nineteen years old and as the Moon moves further away from Earth by 4cm every year, each of us is caused to steadily receive a lessening amount of the neuro-endocrine hormones secreted from our pituitary gland with each passing month. In *How and Why We Age* Dr. Leonard Hayflick explained that all mammals (including humans) and animals that reach a fixed size as adults do indeed age but that the animal species which continue to increase in size indefinitely are non-aging and tend to be more primitive ones like lobsters, tortoises, alligators, and fish such as sturgeon and shark (1994, pg. 232). Author Joseph Panno posed the question - how is it fair that we humans in our "advanced civilization" are required to age in the first place? Panno writes "after all, we have a reasonably resilient immune system; we heal well after injury; we have enzymes that monitor and repair our DNA (deoxyribonucleic acid); and providing that we eat well, our cells have plenty of energy to take care of themselves every day yet, we age with certainty" (2011, pg. xv).

According to Newton's Physics at some point in the future the Moon will stop its recession from Earth and the gravitational relationship that exists between Earth and the Moon will stabilize, meaning that one-day human beings may be able to live for thousands of years - a possibility which might already be built into our genome through the physiological feature known as *organ reserve* - that organs in the human body require only the functioning of a fraction of their specialized cells in order to perform optimally, with the remaining percentage of cells there only as 'back-up' at a time of illness or injury. If this principle is applied to the ovaries (which contain about four-hundred thousand eggs at birth) and if the average woman were eventually able to release just one-tenth of her eggs, she would be able to ovulate about forty-thousand times over the course of nearly three-thousand years.

Dr. Hayflick explained that as a person ages the overall weight of their brain decreases by approximately ten-percent as a result of lost brain cells called *neurons* and less cerebral spinal fluid (Hayflick, 1994, pg. 233). This means that humans might 'cheat' aging if we could discover a clinical way to compensate for the physiological effects of the Lunar-recession and the resulting decrease in the tidal volume of cerebral spinal fluid throughout each Monthly Cranial Respiration Cycle. Clinical life extension might come in the form of carefully timed injections of replenishing cultured cerebral spinal fluid directly into a one's central nervous system, a procedure that could possibly be helpful in the treatment of certain age-related diseases as Dr. Kern pointed out that the brain's *choroid plexii* (which produces cerebral spinal fluid) deteriorates as one ages and that a deficiency in the production of cerebral spinal fluid leads to malnutrition of the brain and neuro-degenerative age-related disorders such as *Parkinson's disease* (a disorder of motor-function) and *Alzheimer's disease* - the most common of the group of disorders known as *Dementia* which is characterized by a loss of brain function that gradually worsens over time (Kern, 2005, pg. 42). Alzheimer's disease and its associated dementias have been linked to several age related cognitive declines, neuron death, disruptions at synapses between neurons, and to characteristic lesions in the brain (alz. org, 2014). Researchers looking for the cause of dementia are investigating both environmental influences and predisposing genetic markers while primarily studying the abnormal quantity of what is called *amyloid plaque* which consists of "chemically sticky" fragments of the *myelin sheath* that insulates neurons (alz.org). Small pieces of the myelin sheath break-off, accumulate, and clump together to block cell-to-cell transmission at synapses and prevent nutrients, oxygen, and waste from moving through the cell, which subsequently dies. This plaque is seen throughout the brain but is more concentrated in the *hippocampus* (which is known to play a key role in forming new memories). Amyloid plaque is present in typical aged brains; however an Alzheimer's affected brain has a considerable amount more as well as ventricles that grow distinctively larger and a *cerebral cortex* that shrinks in size, especially in the hippocampus. Dementia is twice as common in women and most prevalent in individuals with a family history of the disease, longstanding high blood-pressure, and cranial trauma. It is possible that the Lunar-recession is causally related to the larger quantity of amyloid plaque and the observed characteristic deformation within the Alzheimer's affected brain as the resulting decrease in the amount of cerebral spinal fluid inhaled into the central nervous system with each passing Monthly Cranial Respiration Cycle decreases cranial suture joint motion and creates an immense compression on the brain, causing the ventricles to enlarge, myelin sheath to tear, and pieces to break-off and clump together at areas in and around the region of the hippocampus. This means that a woman's two-fold predisposition to dementia may simply stem from the facts that the average female reaches adulthood and stops increasing in height about two-years sooner than the average male thus her brain experiences the effect of the Lunar-recession for longer. And, females are known to have a naturally higher resting blood-pressure than males, leaving their brain tissue more susceptible to the increased compression resulting from the cranial suture motion atrophy brought on by the Lunar-recession. Notwithstanding gender, it is conceivable that a moderate to severe injury to anyone's cranium is likely to cause their cranial suture motion to atrophy prematurely at particular suture segments between impacted cranial bones thus affecting different regions of the cerebral cortex leading to the wide-range of symptoms displayed within the various manifestations of dementia.

In a similar manner the Lunar-recession and the sustained increase in intra-cranial pressure upon one's brain due to cranial suture motion atrophy may have a causal relation to Parkinson's disease - a condition that progresses over time and presents cardinal signs of abnormal immobility of facial muscles, rigidity, involuntary trembling of the limbs, muscle stiffness, lack of movement, and difficulty initiating movement (Abramovitz, 2005, pg. 13). Individuals afflicted with Parkinson's disease typically bend forward into a 'stooped posture' associated with a great difficulty in maintaining balance when walking and in some cases sufferers cannot stop parts or even their entire body from moving (Sharma, 2008, pg. 1). Symptoms often begin on one side and then develop bilaterally. Scientists know that neurons use a chemical transmitter called *dopamine* to control all muscle movement. Parkinson's disease begins when the brain cells that make dopamine known as the *substantia nigra* are slowly malformed and destroyed (Abramovitz, 2005, pg. 29). These specialized brain cells are located immediately above and on either side of the hippocampus. Scientists do not know what causes their destruction and malformation however, as with Alzheimer's disease both environmental and genetic causes are being investigated. Presently, aging is the only known risk factor for a person to contract Parkinson's (Sharma, 2008, pg. 6 & 35). A causal relation between Parkinson's and cranial suture joint motion atrophy brought on by the Lunar-recession may exist in the form of the increased compression on an individual's brain originating primarily from the *occipital bone* at the posterior cranium. This exerts pressure directly onto the posterior brain *or cerebellum* (which controls muscle movement) that is then transferred forward and upward to cause brain-tissue tearing at areas directly above and beside the hippocampus. An anatomical predisposition such as this might stem from a cranial event that occurred at some point earlier in a person's life (even during one's birth) as physician Dr. John Upledger (the founder of Craniosacral Therapy) explained in the book *Your Inner Physician and You* that quite often during its birth a baby's head is hyper-extended and pulled-on in an attempt to hasten the delivery process, a practice which has been known to compress the occipital bone and cause it to grow misaligned or end-up 'blocked' and not moving optimally or not moving at all (1997, pg. 25). Upledger cited several examples of how a longstanding existence of this type of occipital or *temporal bone* blockages throughout one's life could potentially contribute to the dysfunction of their craniosacral system and lead to conditions such as hyperactivity and learning disabilities. He explained that even a mild episode of meningitis occurring as a child could cause the development of scar-tissue and adhesions in one's cranial membranes which can decrease the vitality of their craniosacral-rhythm and cause the obstruction of arteries leading to a reduction in blood-flow and a diminished brain capacity (Ibid). Dr. Upledger asserted that 'it is only when an individual's craniosacral-rhythm is able to find its natural expression around the different fulcrums that *homeostasis* or wellness can take place' (Agustoni, 2008, pg. 248).

The battle against all disease benefits from further investigation into the Monthly and Daily Cranial Respiration Cycles, the Biological Tides Theory, and the possibility that the Moon's gravitational pull regulates comparable cycles within viruses, bacteria, parasites, and certain cancers. Identifying solar and lunar cycles within these pathogens may reveal times of each day and month at which they are more susceptible to treatment, allowing health practitioners to more effectively schedule their treatment-plans.

Space travel and the Physiological Effects of the Lunar-recession

Specific ramifications of the regulating affect that gravity has on each person's overall cranial respiratory health are possible regarding space-travel as Dr. Roberta Bondar (the first neurologist in space) described how the weightlessness of free-fall affects every system of the body and can make astronauts lose their appetite and even become ill. In a Canadian Broadcasting Corporation radio interview with Michael Enwright, Dr. Bondar explained that after living for a prolonged period in microgravity astronauts return to Earth's surface with specific irreversible physiological changes such as blindness (due to increased intra-cranial pressure upon the *optic nerve*), bone-loss, short-term motor dysfunction, and an increase in the presence of amyloid-plaque in their brain-tissue. Dr. Bondar revealed that while aboard the *Space Shuttle Discovery* for eight-days in 1992 there were notable changes to her own menstrual cycle. When questioned about a manned mission to Mars, she stated that she `is not quite sure humans are ready for such a trip (Enwright, 2014).

Presently, there are two privately funded planned manned missions to Mars. One is poised to send a married couple on a fly-around of the red-planet in 2018 and have them return to Earth after more than five-hundred days in microgravity. The other mission intends to send groups of four on one-way trips to land on Mars and establish a permanent human colony there, beginning in 2026. These space pioneers face unprecedented neuro-endocrine hormonal secretion challenges as they move millions of kilometers away from Earth and the regulating affect of the Moon and its gravity. Once on Mars, which only has one-third of Earth's gravity and two tiny moons, it is possible that the astronauts will experience acute cranial suture motion atrophy causing their Monthly Cranial Respiration Cycle and 'three tides' to diminish enough that it leads to their becoming ill.

It might be prudent if manned space exploration focused on missions to worlds that are not as far away from Earth as Mars is, and to ones that are more Earth-like such as the planet that is one position closer to the Sun than Earth (Venus) or the body one position closer to the Sun than Venus (Mercury) - both of which are thought not to possess a moon. Venus has been visited by several Orbiters and Landers which found it to be the 'twin sister' planet to Earth (Moore, 2002, pg. 10), however humans are not likely to colonize Venus as its average surface temperature is four-hundred degrees Celsius, hot enough to melt lead. On the other hand, in 1974 NASA's probe *Mariner-10* revealed that Mercury is almost identical to the furthest place that humans have visited and returned from safely - the Moon (Strom, 1987, pg. 1). Previously, Mercury had been dismissed by many astronomers as the 'least interesting' terrestrial body, due to the fact that its unique proximity to the Sun renders it the body that is most difficult to view, a feature which has surely contributed to the large amount of confusion that has always seemed to surround this elusive body. For instance, early civilizations thought that Mercury was two separate bodies due to the fact that sometimes it appears on one side of the Sun as a second morning star, while at other times it appears on the opposite side of the Sun as a second evening star. When Mercury was visible in the morning the Greeks referred to it as 'Hermes' and in the evening they called it 'Apollo'. The ancient Egyptians called these two bodies 'Set and Horus' while Hindus knew of them as 'Boudhavava and Raulineya'. By the time of Plato most civilizations had recognized that the second morning star and the second evening star were the same thing and, based on the way that it appears to move like a planet in an orderly manner west to east across the horizon, concluded that it was the closest planet to the Sun. Mercury's quick speed is what inspired the Romans to name it after Zeus' quick-winged messenger and their god of commerce - Mercury.

During modern times the confusion over Mercury persisted as astronomers assumed that Mercury always showed the same face toward the Sun. Eventually, it was realized that this is incorrect. Today, scientists are baffled as to why Mercury's orbital eccentricity and inclination are so much larger than all other bodies in the solar-system except Pluto. One possibility is that Venus' gravitational perturbations caused their increase. Further mystery concerning Mercury surrounds the presence of its magnetic field as it was not until the voyage of *Mariner-10* that it was discovered that Mercury is magnetized with a North and South Pole. Since this discovery there has been much debate as to where this magnetic field originates (Lang, 2003). Planetary scientists had always thought that a prerequisite for a planet to generate a magnetic field is through a *dynamo* - the quick rotation of an electrically conductive molten metallic core (Brewer, 1990). Scientists believed that Mercury, dense for its size at 11,900 pounds per square cubic yard, spins too slowly for a dynamo to be capable of sustaining within its dense iron interior and that its core (larger in-proportion to its overall size than the other terrestrial bodies) should have fully cooled and solidified long ago as smaller bodies radiate their heat into space quicker thus losing their dynamos more rapidly (Lang, 2003).

In 2011 NASA's *Mercury* MESSENGER craft became the first to orbit Mercury, making Mercury the new possible 'missing link' which might harbor the clues that reveal how the terrestrial planetary systems came into existence. Although it is believed that Mercury fulfills the criteria currently required for a body to be receive an official planetary designation from the *International Astronomical Union* - it is spherical with a cleared and dominant orbital path around the Sun - astronomers Drs. Thomas Van Flandern and Robert Harrington of the U.S. Naval Observatory jointly explained that the composite picture of Mercury seems more like the description of a moon than it does a planet (Icarus, 1976, pg. 435). The next section of this work combines Drs. Van Flandern and Harrington's characterization of Mercury with Classical Newtonian Mechanics, the Theory of General Relativity and recent data from MESSENGER to move forward the idea that Mercury is the moon to the planet Venus which recessed from Venus in a manner relative to how the Moon is still recessing from Earth.

Mercury is Not a Planet

The Newtonian mechanics displayed within Earth and the Moon's gravitational relationship and their similarity to the Newtonian mechanics displayed between the planets further from the Sun and their moons, make it rational to suggest that planets in this solar system must possess at least one moon *in order to establish a stable orbit at a relative distance from the Sun*. For example, the time it takes *Charon* to orbit Pluto is equal to the amount of time it takes Pluto to rotate once on its axis, so that if astronauts ever do stand on Pluto they would always see the same face of Charon, meaning that Pluto and Charon are locked in *synchronous rotation* as are Earth and the Moon (Goss, 2003, pg. 14), (Cazenave, 1988, pg. 55).

The suggestion that all of the planets in the solar system must possess at least one moon is problematic as it is in direct contrast to the doctrine that both Venus and Mercury do not possess a moon. Yet, Drs. Van Flandern and Harrington's joint paper titled *A Dynamical Investigation of the Conjecture that Mercury is an Escaped Satellite of Venus* (1976) examined whether Mercury originally formed as a satellite to Venus and then migrated to its present distance from the Sun. In this paper the two astronomers explained that the mechanism for the loss of Venus' angular momentum derives from the likelihood that Venus originally had a close satellite that was similar to Earth's Moon. They also pointed out that one-hemisphere of Mercury and the Moon is more cratered while the other hemisphere of each body is smooth or *mare* and that the possibility for this asymmetry in the case of the Moon, is that the Moon was shielded by Earth from meteoritic impacts that occurred at a time when the two were still quite close together. Drs. Van Flandern and Harrington explained that the notable asymmetry on Mercury's surface is totally inexplicable, unless Mercury was a planetary satellite and closer to Venus at the time of its formation and then moved further away (Icarus, 1976, pg. 435).

The authors began researching the above mentioned paper shortly after *Mariner-10* had revealed for the first time that Mercury and the Moon are not only similar to each other in size but almost identical to each other in their surface presentation (Moore, 2002, pg. 27), (Whipple, 1981, pg. 178). Both are heavily cratered and barren, both contain huge multi-ringed basins, and evidence of ancient lava-flows while each is covered by a loose porous layer of fine grained grey dust called *regolith* (Lynch, 1999, pg. 37), (Timelife, 1990, pg. 61). Mercury's regolith is texturally similar to and possesses the bearing strength of lunar soil such that, if astronauts were ever to walk on Mercury, it would be difficult to distinguish their footprints from those on the Moon (Burgess & Dunne, 1978, pg. 85). Mercury and the Moon's surfaces are the most ancient and inactive in the inner solar system (Brewer, 1992, pg. 10) and both display evidence of an episode of planet-wide surface melting near the end of their first or *early period* of heavy meteoric bombardment, when each body was similarly pockmarked by the same barrage of projectiles early in the solar system's history (Burgess & Dunne, 1978, pg. 23). Mercury and the Moon then simultaneously experienced a second or *late period* of heavy meteoric bombardment yet have since remained geologically inactive and relatively un-evolved (Lang, 2003, pg. 49), though it is evident that erosional spin and meteoroid impacts have caused a degree of crater demolition on the surfaces of both bodies. This demolition is easily measured as neither Mercury nor the Moon experiences any weather to erode their surface (Morrison, 1993, pg. 72), (Fischer, 1987, pg. 83). The ubiquitous meteorite impact craters on Mercury are similar to the meteorite impact craters on the Moon as they possess secondary craters, crater chains, inner-walls, ejected material, centralized peaks, and great circle alignments that closely resemble the *ray craters* on the Moon (which are brighter than the adjacent area and surrounded by what look like 'rays' created by material that is ejected upon impact) (Chapman, Mathews & Vilas, 1981, pg. 132), (Burgess & Dunn, 1978, pg. 126). Additionally, Mercury and the Moon are the only two of the five inner solar bodies that lack an appreciable atmosphere, yet they present comparable rare and trace tenuous conditions inside the vacuum of space, with Mercury demonstrating the same sort of temporary sodium and potassium clouds detected on the Moon (Lunine, 1999, pg. 14), (Steel, 2000, pg. 35). Neither possesses water but ice has reportedly been found in deep polar craters on each body (Emiliani, 1992, pg. 142). They also share both polar and photo-metric properties as the relative brightness and the way sunlight reflects off of the surface of each body are identical (Chapman, Matthews & Vilas, 1988, pg. 37), (Burgess & Dunne, 1978, pg. 77).

On the surface of both the Moon and Mercury there is a large distinct three-ringed basin which clearly resembles a target or 'bull's-eye' intersected slightly off-center by its *terminator* which is the line that divides the lit and unlit hemispheres. Scientists refer to the three-ringed basin on Mercury, the largest in the solar system and the most prominent feature on the Mercurian surface, as the *Caloris Basin*. Caloris is Latin for heat and the basin is so named because it is located at Mercury's *sub-solar* point or the location on Mercury's surface which is almost directly under the Sun every other time Mercury is at *perihelion* or its closest approach to the Sun. The Lunar counterpart to the Caloris Basin is the *Orientale basin* which is located at the exact center of the Moon's forward facing surface (perpendicular to Earth) at the Moon's most westerly limb. These two basins are considered to be the freshest and largest in the solar system, proportionate to each other in size, and many stratigraphic and structural similarities have led scientists to the certainty that Caloris and Orientale formed of an identical process at approximately the same time, between 3.8 and 3.9 billion years ago (MacCauley, 1977 pg. 242).

Directly opposite to the center of the Orientale and Caloris basins on the other side of Mercury and the Moon, respectively, are areas known as the *antipodes* where there is an atypical severely disrupted surface known as *weird terrain* consisting of hills and lineations on which the rims of craters are broken and dissected (Audouze & Israel, 1988, pg. 69), (Moore, 2007, pg. 9). The weird terrain at the antipode to the Caloris basin on Mercury, similar to but larger in extent to the corresponding area on the Moon, is thought to be the result of seismic energy from the meteorite impact that is believed to have created the Caloris basin (Taylor, 2001, pg. 41). Scientists believe that the lunar weird terrain is also the result of similar seismic energy from the meteorite impact that is believed to have created the Orientale basin (Moore, 2006, pg. 10). There is conjecture that the respective impacts which formed the two large multi-ringed basins sent shock waves through to the other side of Mercury and the Moon, causing the formation of the weird terrain (Burgess & Dunne, 1978, pg. 81).

Scarps or cracks in the Mercurian surface point to compressional forces which indicate that since the time of the Caloris forming event Mercury's radius has gradually contracted by approximately 1.5 – 2 kilometers (Bakich, 2000, pg. 101), (Thomas, 1988, pg. 69). Stressed induced *rilles* on the Moon and their similarity to scarps on Mercury suggest that the lunar crust has similarly shrunk since the Orientale forming event (Thomas, 1988, pg. 110), (Lang, 2003, pg. 201). At present, Mercury and the Moon have dimensions, masses and densities which are similar to those of Jupiter's four *Galilean* satellites. Mercury is comparable to both *Ganymede* and *Callisto* while the Moon is comparable to *Io* and *Europa* (Thomas, 1988. pg. 111). Mercury and the Moon each have an iron metallic core which is known to generate a small magnetic field (Grinspoon, 1997, pg. 164), (Audouze & Israel, 1988, pg. 176). They are the only two of the five inner solar bodies that have a rotational axis that is perpendicular to its orbital-plane (Sprague & Strom, 2003, pg. 43). In 1975 NASA convened a conference in Houston, Texas to study the similarities between Mercury and the Moon. Attending scientists reached a consensus that the geological history existing on the Mercurian surface indicates that *the far reaching similarities between Mercury and the Moon suggest that the processes that shaped the two bodies were the same* (Kopal, 1979, pg. 106), These similarities become even more explicable as the following physical similarities between Venus and Earth and the Newtonian mechanics displayed within their motions, reveal that they are 'twin sister' planets and that Venus must possess a moon which is comparable to Earth's moon.

Venus and Earth as 'Twin Sister' Planets

Prior to Drs. Van Flandern and Harrington's investigation into the conjecture that Mercury is an escaped satellite of Venus, the astronomical community had confirmed that in size, density, composition, and distance from the Sun, Venus is almost identical to Earth (Cooper, 1993, intro), (Grinspoon, 1997, pg. 196). Venus' general appearance is like Earth's and its diameter is 95% of Earth's diameter. The two planets are about the same age and Venus' average distance from the Sun is only 25% closer than Earth (Fisher, 1987, pg. 88), (Moore, 2002, pgs. 12 & 36). Venus' total volume is 86% of Earth's while it appears to have a metallic core and an internal arrangement similar to Earth's interior (Burgess, 1985, pg. 138). Venus' surface gravity is 91% of Earth's and its density is 5:2 times the density of water, close to Earth's 5:5 times the density of water (Brewer, 1992, pg. 21). Venus is a "volcanically active sister to Earth" (Hartman, 1993, pg. 22) with similar tectonic plates and migrating continental masses (Burgess, 1985, pg. 138). There are two tectonic masses on Venus which are comparable in size to the United States (Whipple, 1981, pg. 185) while one highland area on Venus is equivalent in size and elevation to Africa and another is similar to Australia (Alter, Cleminshaw & Phillips, 1983, pg. 131).

Venus, like Earth, has long mountain ranges, fault-lines, valleys, plains, plateaus, deep chasms and mountains with similar average heights as Earth's (Kerrod, 2002, pg. 21) while at the same time Venus has low rolling plains that cover about two-thirds of the planet's surface making them comparable to Earth's ocean basins which cover about two-thirds of Earth's surface. The crystal-rocks on Venus are similar to Earth's granite and magmatic rocks (Cambridge Atlas, 1994, pg. 85). Venus has an atmosphere as substantial and actively complex as Earth's (Wagner, 1991, pg. 180) (Israel, 1988, pg. 70). Comparable amounts of gaseous nitrogen and carbon exist on the two planets while there is known to be an amount of *atomic hydrogen* in the upper atmosphere of Venus which is comparable with that in the upper atmosphere of Earth (Burgess, 1985, pg. 25), leading geologists to believe that each of the sister planets developed an atmosphere with gases that were expelled from their interiors (Colin & Luhmann et al, 2002). In fact, both Venus and Earth are known to possess a dense *hydrogen corona* through which hydrogen atoms are emitted into space (Grinspoon, 1997, pg. 107). Venus is covered by clouds that display frequent lightning flashes sometimes as many as twenty-five per second at an altitude between one and five-kilometers above the surface (Wagner, 1991, pg. 132). Scientists are in full agreement that Venus was once wet and may have had broad oceans at one time (Colin & Luhmann et al, 2002). Today there are only ten-centimeters of hot *heavy water* vapor remaining in Venus' atmosphere which is about ninety-percent carbon-dioxide. On Earth, there is a matching percentage of carbon-dioxide locked away in carbonate rock. Astronomers have observed that on all

Earth-like planets, water and carbon-dioxide are the most prominent volatile substances.

To assist in the understanding of how the similarities between Venus and Earth and the Newtonian Mechanics displayed within their respective relationship with the Sun dictate that Venus must possess Mercury as its moon, consider that if the Moon did not exist Earth would not orbit the Sun within the *habitable zone* (the region around the Sun at a distance where water can exist in all-three states and life has formed) meaning that Earth's *sidereal period* or the time required for it to travel completely around the circumference of the Sun once (365.25 Earth-days) would be entirely different (Daniels, 2005, pg. 102). Next, realize that what scientists consider to be Venus' sidereal period of 224.7 Earth-days is improperly measured and requires re-assessment in order to reflect one that is more similar to Earth's observed sidereal period. This declaration is affirmed within Classical Newtonian Mechanics as all objects that are moving under the influence of a centrally directed force or *nucleus* varying with the square of the distance - orbit in an elliptical path that *does not vary with time.* In order for this Classical Newtonian Mechanical law to be observed within an object's orbital motion, the observed object must at the end of its orbit return to the exact same spot in space at which its orbit began and then repeat the identical orbital path without any difference. However, scientists know that neither Earth nor Venus is observed to return to the exact same point in space at what is considered to be the end of their sidereal periods 365.25 Earth-days and 224.7 Earth-days, respectively (what astronomers usually refer to as their orbital periods) (Maor, 2000, pg 59). Neither planet officially fulfills the Newtonian mechanical law until the passage of 2922 Earth-days, when they not only return to the exact same spot in space relative to one another, the Sun, and the stars but they come to their closest possible approach to each other while the same face of Venus that was turned toward Earth at the period's beginning is the same face of Venus that is turned toward Earth at the period's ending. Each planet is then observed to repeat the exact same orbital path throughout the following 2922 Earth-day period (Elkins-Tanton 2006, pg. 71). This means that the 'twin sister' planets are fulfilling the Newtonian mechanical law defining body's orbit, and orbiting the Sun in an identical amount of time, 2922 Earth-days. Except, due to the fact that Venus is approximately twenty-five percent closer to the Sun than Earth, Venus travels in a tighter almost circular orbital path, at a speed which is approximately five-kilometers per hour faster than Earth, meaning that in order for Venus to express a relationship with the Sun that is relatively similar to her 'twin sister' planet's relationship with the Sun, Venus is required to lap the Sun more times than Earth does during each of the 2922 Earth-day periods. By dividing 2922 Earth-days by Venus' sidereal period of 224.7 Earth-days and by dividing 2922 by Earth's sidereal period of 365.25 Earth-days it can be determined that during 2922 Earth-days, Venus completes thirteen laps of the Sun and Earth laps the Sun eight times, precisely. At this ratio it can be calculated that Venus is traveling 1.625 times around the circumference of the Sun during the same amount of time that Earth travels in 1.0 of its trips around the Sun. This corroborates that Venus' sidereal period of 224.7 Earth-days (as mentioned above) requires re-measuring in order to account for this extra 0.625 of a lap of the Sun difference. Multiplying 1.625 by 224.7 Earth-days provides that when the extra 0.625 distance that Venus must travel is factored into Venus' motion, Venus' sidereal period is more properly calculated at 365.21 Earth-days, almost identical to Earth's 365.25 day sidereal period and, given that the Moon is a significant factor in the length of Earth's sidereal period, it can only be construed that Venus possesses Mercury as its supplementary moon - an inference based not solely on Mercury's proximity and its lunar-like atmospheric and geological features internally and externally, but also on the way that Mercury co-orbits Venus and the Sun in a manner that is similar to how the Moon co-orbits Earth and the Sun. This is observed as Mercury laps the Sun 2.5 times, precisely, each time that Venus laps the Sun once, meaning that there are three superior and three inferior *conjunctions* or six alignments of Venus and Mercury each time that Venus laps the Sun once (one quarter of the amount of conjunctions that occur between Earth and the Moon every time Earth laps the Sun once). At the same time, Mercury rotates one and a half times on its axis during one of its laps of the Sun. This feature is known as *spin-orbit coupling* (Alter, Cleminshaw & Phillips, 1983, pg. 123) and not only does it cause Mercury's day to be longer than its year, it is the reason why it takes two of Mercury's trips around the Sun for the Sun to return to the same given position in Mercury's sky as it was the previous day (Taylor, 2001, pg. 40). More significantly, the coupling leads to the alternation between longitudes 0 and 180 on opposite sides of Mercury as the *hot-poles* or closest points to the Sun, meaning that Mercury experiences two separate perihelions during one of its laps around the Sun. In turn, Mercury experiences two separate *aphelions* (furthest point from the Sun) or *warm-poles* each time it laps the Sun once. These occur alternately between the hot-poles at longitudes 90 and 270 (Croce, 2005, pg. 24) (Elkins-Tanton, 2006, pg. 106). Technically, Mercury's two alternating perihelions and aphelions serve as alternating apogees and perigees to Venus, such that during Venus' newly re-measured sidereal period of 365.21 days, Mercury functionally orbits Venus six times (half of the number of times that the Moon orbits Earth during one of Earth's 365.25 day sidereal periods). Additionally, Mercury crosses Venus' orbital plane or *ecliptic* at 4.0 degrees, creating two *nodes* (one ascending and one descending) in Mercury's path that intersect Venus' ecliptic similarly to how the Moon's ascending and descending nodes intersect Earth's ecliptic at 5.9 degrees.

The recognition of Mercury as Venus' moon offers an explanation for a well known anomaly present within Mercury's orbital motion. This famous incongruence occurs at Mercury's perihelion where the small terrestrial body reaches a speed of fifty-eight kilometers per-second and makes it to as close as a mere forty-six million kilometers from the Sun. In fact, Mercury gets so close to the Sun at its perihelions that it is believed the Newtonian equation for gravity incorrectly predicts Mercury's orbital path as the expected *transits* of Mercury across the face of the Sun start as much as a day or at the very least several hours late. This means that each of Mercury's subsequent perihelions does not occur where it is predicted to by Newtonian Mechanics. Instead, Mercury's closest point to the Sun *precesses* or occurs slightly ahead of the location of the previous perihelion, causing Mercury's entire orbit to rotate in space in a distinct rosette pattern. This rotation progresses at such a slow rate that it takes 244,000 years for Mercury to trace out the complete pattern around the Sun and then return to its exact starting point. According to Newtonian Mechanics it should only take 227,000 years for Mercury to trace out the complete pattern. It has been determined that the position of Mercury's perihelion moves 5,600 *seconds of arc* each century and, although it is well known that other planets regularly perturb Mercury and cause its path to deviate, this amount is forty-three seconds of arc per-century more than can be accounted for using Newton's physics - *the only known inconsistency within all of Newtonian Mechanics*. When this irregularity in Mercury's motion was discovered, astronomers were fully convinced that Mercury's incongruous motion was due to the Newtonian pull of a 'not yet discovered Mercury perturbing planet' closer to the Sun than Mercury. The discrepancy is believed to have been explained by Albert Einstein's Theory of General Relativity as it later replaced Newton's equation for gravity with *curved space-time* and the realization that *bodies closer to the Sun behave differently than bodies further from the Sun* (the first known great success of General Relativity). Understanding that Mercury is the moon to Venus provides that both General Relativity and Classical Newtonian Mechanics jointly contribute to the explanation of the incongruence within Mercury's motion once Mercury's perihelions to the Sun are recognized as its apogees to Venus precessing in a *prograde* or clockwise direction in relation to Venus' *retrograde* or counter-clockwise axial rotation relatively to how the Moon's perigee precesses in a retrograde direction in relation to Earth's prograde axial rotation.

The understanding that Mercury is Venus' moon provides us with new clues necessary for deciphering the exact process of how the inner terrestrial bodies formed. The most obvious of these clues are certainly the overall similarities between the two largest and most well preserved basins in the solar system (the Orientale and Caloris basins) on the Moon and Mercury, respectively. As outlined above, the presently accepted model of the formation of these two essentially identical three-ringed basins requires the supposition that they are the result of random meteorites that impacted Mercury and the Moon almost simultaneously, with enough force to breach the mantels of Mercury and the Moon and cause a proportionate amount of lava to flow out of their interiors and up to their surfaces, but with little enough force as to not entirely obliterate either Mercury or the Moon in the process. The accepted impactor model also supposes that the Orientale and Caloris forming impacts occurred at locations where each basin's center is slightly offset to the west of both Mercury and the Moon's terminator and that each impact similarly sent shock-waves of seismic energy through and around both bodies creating the weird terrain at the antipodes of the two multi-ringed basins. Contrastingly, astronomer Dr. Linda Elkins-Tanton explained in *The Sun, Mercury, And Venus* that ``even though it is understandable to think that shock-waves from these hypothetical impacts traveled toward the opposite hemisphere of each body, there is no adequate theory of the effect that giant impactors have at the antipode of their impact point`` (Elkins-Tanton, 2006, pg. 100). Certainly, the overwhelming resemblance of the Orientale and Caloris basins and the weird terrain at their antipodes are too similar to each other to be the result of random meteorite impacts as the impact theory has proposed.

Instead, consider the preliminary data sent from MESSENGER which, before it ran out of fuel and was programmed to impact the Mercurian surface, detected volcanic vents circling the three margins of the Caloris Basin. Astronomer Dr. James Head explained that "MESSENGER's detection of these volcanic vents in and around Caloris answers an age old question amongst planetary scientists; that the smooth plains in the inner Caloris basin were caused by erupting lava" (nasa.gov, 2010). Concerning the Moon's Orientale basin, Dr. Head similarly explained in *Morphology and Distribution of Volcanic Vents in the Orientale Basin* that "the ubiquity of the volcanic vents found in and around Orientale offers substantial clues to the causal relationship between basin interiors and mare volcanism" (2011, pg. 3).

Dr. Head's findings suggest that the Caloris and Orientale basins are not the result of meteorite impacts as has been presumed, rather they are the geological markings of volcanic material that *exited* both proto-mercury and the proto-moon simultaneously through a series of tidally-induced volcanic eruptions which occurred at about the time that the proto-sun's gravitational compression upon them increased during their early period of heavy meteoric bombardment, over four-billion years ago. The initial eruptions caused the outer rings of each of the respective basins and surface-wide melting as lava began to ooze up to the surfaces of the two proto-moons creating their *Intercrater plains*. The second volcanic eruptions were larger as much denser material from deeper within proto-mercury and the proto-moon exited through Caloris and Orientale, only a few hundred million years before the third and most violent eruptions spewed the deepest and densest portion of their molten iron-nickel cores out into the primordial solar nebula, where they coalesced and *accreted* into the two large planetesimals proto-venus and proto-earth. During this process a barrage of meteors showered one hemisphere of proto-mercury and the proto-moon in a second or late period of heavy meteoric bombardment which ended approximately three-billion years ago as proto-venus and proto-earth's mantels *mass-fractionated* into the spherical 'twin sister' planets Venus and Earth. The immense increase in the size of the twin sister planets caused each planet's axial rotation to begin to *brake* and the conservation of that energy to transfer onto Mercury and the Moon, respectively, both of which obliged with an increase in orbital speed thus moving further away and entering into space-time closer to the Sun and deeper within its gravity. Tidal forces caused magma to seep up to the surfaces of Mercury and the Moon through fissures and intrusions in their crusts bringing on the widespread basaltic lava flows which created Mercury's *smooth plains* and the *Maria* on the Moon's Earth-facing surface. As their cores cooled, both Mercury and the Moon shrank creating the scarps and rilles in their crusts and the weird terrain at the antipodes of the Caloris and Orientale basins.

The Atomic Planets

Considering Mercury as Venus' moon mandates that each of the first eight planets in the solar system - *Venus, Earth, Mars, Jupiter, Saturn, Uranus, Neptune, and Pluto* must possess at least one moon. In a planetary sense, this is analogous to how all atomic nuclei must possess at least one electron, such that the solar system has atomic structure - eight bodies orbiting one central nucleus - the Sun. An atom's nucleus contains almost all of an atom's mass while the Sun comprises ninety-nine percent of all the matter in the solar system. Contextually, the solar system is structured as a *planetary atom* - where the force of attraction between an electron and a proton in the nucleus, equally matches the outward pressure resulting from the movement of the electrons (Helmenstine, 2014). For stars like our Sun to attain stability, the pressure from the mass of the star's outer layers upon the core must equally match the outward pressure created by the temperature of the *nuclear fusion* taking place within the star's core (Beatty-Chaikin, 1990, pg. 15). In other words, all of the matter in this *heliosphere* (down to the last atom) is required to ensure the Sun's stability.

The Sun and its eight orbiting planetary systems resemble the structure of an atom of the third most abundant element within the Sun, the *non-metal – oxygen* which consists of atoms that possess eight orbiting *electrons*. Astronomers know that all major satellites originated through the process that created the solar system, and that moons and planets are bi-products of the formation of stars (Cazanave, 1988, pg. 58). Thus, as the Sun formed, its retinue of moons gave birth to eight planets in order to fill its eight outer energy-levels (known as a *stable octet*) with atomically structured orbiting systems that are *covalently bonded* to itself and in at least one instance, to each other. Covalent bonding is a type of chemical attachment that arises from the sharing of one or more electron pairs between atoms of the non-metal elements, for the purpose of becoming *stable* - the driving force in chemical bonding. Once an atom becomes stable it will no longer gain, lose, or share electrons. Oxygen atoms form more than one covalent bond when filling their eight *orbitals* or shells, the first-two of which are always filled by hydrogen atoms. In the case of our solar system, these would be the first two inner-planets Venus and Earth, as the structure of Venus with one moon (Mercury) and Earth with its Moon, are of the simplest of all functioning structures (one body orbiting around one nucleus), an atom of the first element on the Periodic Table of Elements and the lightest and most abundant element not only in the Sun but in the entire universe - the non-metal - *hydrogen*. Mercury and the Moon qualify as a hydrogen atom 'electron' due to the fact that they respectively orbit Venus and Earth in functional ellipses while spinning on their axis and generating a small magnetic field. Hydrogen electrons also orbit their nucleus in an ellipse while generating a small magnetic field. Thus, the arrangement of Venus and Mercury alongside Earth and the Moon resembles that of two hydrogen atoms covalently bonded-to and orbiting the nucleus of an oxygen atom (the Sun) to form an enormous water molecule that could only be related to the origin of Earth's water and how it is believed to have arrived here partially from Venus (Fritz, 2002, pg. 19) as it is known that Venus emits atoms of *deuterium* (a heavier isotope of hydrogen) directly into space that combine with the hydrogen atoms emitted by Earth to form hydrogen-gas molecules which bond with oxygen atoms emitted by the Sun to form water molecules that are spun-out into space where they condense and exist in all three states of matter on the only planet in the *solar habitable zone* - Earth.

Mars, Ceres, and the Asteroid belt

Atomic planetary structure suggests that Mars only has two minuscule moons *Phobos & Deimos* (compared to Venus and Earth's large satellites) due to the fact that the Martian system is arranged as an atom of the second most abundant element in the Sun, the second element on the Periodic Table of Elements, and the non-metal element that is comprised of atoms with two electrons - *helium* - a colorless, odorless, non-flammable, *inert* gas used to inflate airships and balloons, cryogenic research, scuba-diving, and for filling incandescent lamps. The Sun is a result of hydrogen atoms being fused together to become helium.

Beyond Mars is what is known as the *asteroid belt* which contains billions of irregular shaped rocky bodies known as *asteroids*. NASA's *Robotic Dawn* spacecraft recently arrived at the largest asteroid in the belt called *Ceres* which is an almost spherical body with a solid core, icy mantle, and an internal ocean of liquid water under its surface, which is probably a mixture of water-frost and ice with various hydrated carbonate minerals, such as clay, mixed-in (NASA. gov). Even though some astronomers refer to Ceres as the only *dwarf planet* in the inner solar system it does not possess a moon so cannot be considered a planet. However, in January 2014 streams of out-gassing water vapor were detected from shiny regions on Ceres, a feature which is more indicative of a comet.

Jupiter

Past Mars, Ceres, and the asteroid belt is the largest planet in the solar system, Jupiter, with its group of sixteen moons (Bakich, 2000, pg. 228) structured as an atom of the non-metal element which possesses sixteen electrons - *sulfur*. It is not surprising then that Jupiter injects its closest moon Io with pure sulfur through a five-million amps electrical current while at the same time gravitationally squeezing Io. In response, Io continuously vulcanizes molten sulfur and *sulfur dioxide* directly into space from liquid oceans of pure sulfur beneath its surface (Mason, 1988, pg. 181). Sulfur is essential to life and thirteenth in abundance within Earth's crustal composition, a constituent of proteins, fats, bodily fluids, skeletal minerals, drugs and most fertilizers. Chemically, sulfur resembles oxygen and when melted is a straw colored liquid that flows as easily as water. At temperatures between 160-195 degrees Celsius, its viscosity decreases dramatically as it becomes one-hundred thousand times thicker. When sulfur reaches two-hundred degrees Celsius it suddenly turns dark red. Interestingly, within Jupiter's layered atmosphere there is a red colored steady anticyclone storm known as the *Great Red Spot* which, along with the fifteen major gas *belts* and *zones* that construct Jupiter's layered atmosphere, likely corresponds to one of Jupiter's sixteen moons. This is observable as the different bands and zones and the Great Red Spot (all of which are many times the mass of Earth) cannot condense together and occur in the same altitudes from year to year. Occasionally one of them become fainter, narrower, or disappear altogether but then always reappears within the standard layered pattern. Each adjacent layer flows at a different speed and in the opposite direction (Mercury Mag., 2002, Vol.31.1) while the dark bands (known as belts) are regions of descending sulfur and phosphorus based gases and the light bands (known as zones) are raising gases such as hydrogen, helium, and small amounts of heavier elements such as methane and ammonia. Correspondingly, Jupiter's moon system possesses similar orbital regularity as some moons have prograde orbits while others are retrograde. Each time Io orbits Jupiter four-times, Europa orbits Jupiter twice while Ganymede orbits the gas giant once. When Europa and Ganymede are in conjunction with Jupiter, Io is always 180 degrees away from the direction of their conjunction, a relationship known as *Leplace's Relation*. This relation partially drives volcanism on Io and is the reason that the Galilean family of moons is considered a miniature solar system (Audouze & Falque, 1994, pg. 59). Leplace's Relation may also allow the Jovian atmosphere to separate molecules of solar debris that impacts Jupiter and then sort and transfer the various atoms to corresponding Jovian moons from where they are then emitted out into space. An example of this molecular separation occurs when Jupiter is impacted by a comet. In 1993 the comet *Shoemaker-Levy 9* passed close to Jupiter and was shredded into twenty-one fragments by the giant planet's gravity. On the comets' subsequent visit in 1994, the fragments crashed spectacularly into Jupiter and were swallowed-up by the gas giant's atmosphere (Bakich, 2000). Astronomers can only speculate what happened to the dust, water ice, and gasses that surrounded the comet's fragmented core. Could it be that after this material entered Jupiter's atmosphere its molecular structure was separated and the different atoms distributed to the various Jovian moons via *flux-tubes, plasma streams,* or *torii*? What happened to the twenty or so solid iron nickel core fragments that are estimated to have ranged in size from 350 meters to 2 kilometers in diameter? Were they ejected back into space to assume Jovian orbits? This would account for the origin of Jupiter's *Trojan and Greek Irregular Satellites* that are smaller than Jupiter's smallest official moon *Thea* and have gathered on either side of the gas giant. Irregular satellites have three radii and are held together by atomic forces rather than by their own gravity meaning that they are not spherical (Comins, 1993).

Saturn

Saturn along with its eighteen official moons (Bakich, 2000, pg. 253) resembles an atom of *argon* which possesses eighteen orbiting electrons. Argon is a non-metal element that rates third highest on the list of gasses in air. It is the most abundant *noble gas* and was the first noble gas to have been discovered. The apparent atomic arrangement of the Saturnian system presents a clear example of how the formation of the planetary systems involved the process of covalent bonding as astronomers believe that Saturn's moon *Phoebe*, unlike any of Saturn's natural satellites, was captured by Saturn's gravity shortly after Saturn and its group of natural satellites formed. This belief stems from the fact that Phoebe's non-synchronous axial rotation and retrograde orbit take this tiny little moon three-times further from Saturn than the seventeen closer, larger, and prograde orbiting moons (Fisher, 1987, pg. 111). Phoebe's size, capture, and distant orbit conform to the theory of covalently bonded planetary systems in how the element immediately before argon on the Periodic Table of Elements (with seventeen electrons) is the halogen - *chlorine*. Halogens tend to gain electrons and a chlorine atom can catch a *neutrino* (a tiny sub-atomic particle emitted by the Sun) to gain an electron and become an argon atom, meaning that the Saturnian system may have originally formed resembling a chlorine atom (with seventeen close moons) that managed to catch Phoebe and stabilize its octet, achieving the structure of an argon atom. It would also mean that Phoebe is not only the largest neutrino in the solar-system but what is referred to in chemistry as a *valence electron*. Valence electrons always reside in the outermost electron level furthest away from and held most loosely by the nucleus. They are the electrons that are lost, gained, or shared in covalent bonding chemical reactions.

Uranus and Neptune

Uranus and its fifteen moons (Walsh-Sheppard, 1994, pg. 58.) are structured as an atom of the non-metal element *phosphorus*, but in the form of a *phosphate* (a combination of phosphorus and oxygen). Phosphorus is a highly important constituent of DNA - the molecule that is known to encode genetic information in animals and plants. Soil without phosphorus is considered to be barren as this element is an essential component of plant- cell protoplasm and nervous tissue and bones in all animals including humans (Emsley, 2000, pg. 1). The name Phosphorus is from the Greek word that means body which glows in the dark. This element resembles nitrogen which is the element located immediately superior to it on the Periodic Table of Elements (Swertka, 1996, pg. 64) explaining why Uranus and Neptune, the planet which is immediately superior to Uranus, are considered to be the `distant-twins` as Neptune (with seven close original moons) is only a slightly bigger version of Uranus (Copen, Dobbins & Parker, 1988, pg. iii) and likely the largest *nitrogen* atom in the solar system as nitrogen atoms originally possess seven close orbiting electrons. This would explain why Neptune's largest moon *Triton* vulcanizes pure liquid nitrogen directly into space from geysers at its surface and why it is observed to have a pink-hue that is thought to be from thin nitrogen clouds and an evaporating layer of nitrogen ice, only a few kilometers above its surface (Hunt & Moore, 1994, pg. 68), (Bakich, 2000, pg. 89). Nitrogen is a non-metal, eighty percent of air, and one of the most abundant elements in all the cosmos (Heiserman, 1991, pg. 27). It is essential for life and a main constituent of DNA. Nitrogen is also a noble gas, so similar to oxygen and argon atoms, a nitrogen atom must always stabilize its octet. Thus, it is not surprising that Neptune possesses a distant captured eighth moon *Nereid* that, like Saturn's Phoebe, is believed to have been acquired only sometime after its primary planet and the natural satellites formed (Hunt & Moore, 1994, pg. 59). Nereid's peculiar comet-like orbit takes it to well over nine and a half million kilometers away from Neptune, indicating that the Neptunian system is ultimately structured as an atom of the odd electron molecule *nitric-oxide* – which is an active neurotransmitter that is an agent in immunology and the destruction of malignant tumors. The human genome contains three different genes which encode nitric-oxide syntases. It is found in neurons, macrophages and in the lumen of blood vessels.

The Kuiper belt and Pluto

In addition to Nereid, Neptune is orbited by as many as ten spherical bodies with names such as *Sedna, Makemake, Haumea,* and *Eris* known as the *Trans-Neptunian bodies* that travel through the orbit of Neptune but reach distances which range from three to ten-times that of Nereid. Scientists explain that these bodies are technically from the *Kuiper belt* – the region of space beyond Pluto which is comprised of billions of pieces of remnants left over from the formation of the solar system. These Trans Neptunian Bodies are likely the larger of the irregularly shaped *Kuiper Belt Objects* that have over time gathered enough molecules of the lighter frozen volatile gases called 'ices' (such as methane and ammonia) to allow them to become large enough to develop nearly spherical shapes and *long term* seemingly stable orbits. Alternatively, the existence of the Trans Neptunian Bodies could mean that Neptune is somehow involved in forming other molecules through the sharing of moons with the eighth planet from the Sun, the *dwarf-planet* Pluto. This would explain why Pluto and Neptune's orbits intersect each other at the spot in space where they switch as furthest planet from the Sun.

Pluto is known to possess Charon and two sets of smaller distant captured moons. The first set *Nix* and *Hydra* travelHH in circular orbits more than twice as far from Pluto as Charon and on the same plane as Charon suggesting that they were captured only a short time after Charon and Pluto formed. The second set of small captured moons named *Kerberos* and *Styx,* were discovered in 2012 at distances even further than Nix and Hydra, out toward what is called the *scattered disc*. Pluto's five known moons indicate that the Plutonian system has formed as an atom of *boron* which is produced entirely by cosmic-ray *spallation* and a low-abundance element in both Earth's crust and the solar system. This element is essential to life as it plays a role in strengthening the walls of plant cells, making it necessary in soil on Earth. Boron is classified as a *metalloid* but it is similar to the group of non-metals in its ability to form stable covalently bonded molecular networks.

With the understanding that Mercury is not a planet but the moon to Venus, the planetary systems' apparent atomic structure evidences that the unique relationship displayed in which the solar system's arrangement into broad categories of different densities: rocky, gaseous, icy, and distance from the Sun relied directly on elemental chemistry - such that the solar system itself and each planetary system within it resembles an atom of one of seven elements with properties of the periodic group of non-metals, covalently bonded to (something only non-metals and boron can do) the nucleus of an oxygen atom - the Sun. Such an arrangement not only ensures the Sun's stability, it permits moons and their planets to emit their respective atoms into space where the solar wind stirs them to combine and form the molecules that give rise to the solar atmosphere. More than eighty different types of compounds have been found in space, suggesting that the building blocks of life may be a feature of the universe, and not only of Earth. These eight elements; oxygen, hydrogen, helium, sulfur, argon, phosphorus, nitrogen, and boron are the most abundant on Earth and the most common in the universe. They are the essential ingredients for creating and maintaining the only atmosphere to develop life, Earth's.

There is clearly the potential for another collection of planetesimals to evolve into our solar system's present arrangement (Boss, 1998, pg. 45). With this possibility and the understanding that the Sun is merely an ordinary *dwarf-star*, identical to thousands of millions in our galaxy alone, and at an average distance (eight-thousand parsecs) from the center of our late-type metal-rich spiral *Milky Way Galaxy*, there is reason for us to consider that the building-blocks of our Sun, the moons, and their planets could behave in a similar way again during the formation of other identical stars within the *galactic habitable-zone*. We may speculate that the atomic solar archetype detailed above has formed around countless Sun-like stars and that on the second planet from at least one of those stars there may be 'life' similar to here on Earth.

Bibliography

Abramovitz, M. (2005). Parkinson's disease: Lucent Books

Agustoni, D. (2008). Craniosacral Rhythm, Churchill Livingston Elsevier

Alter, Cleminshaw, Phillips (1983). Pictorial Astronomy, 5th Revised Edition:

Alzheimer's Association website (www.alz.org). - search Facts and Figures:

Asimov, I. (1996). Earth's Twin, The planet Venus: Garth Stevens Publishing

Audouze, J., Falque, J., Israel, G., (1994). The Cambridge Atlas of Astronomy: Cambridge University Press

Audouze, J., Israel, G. (1988). The Cambridge Atlas of Astronomy: Cambridge University Press

Bakich, M. (2000). The Cambridge Planetary Handbook: Cambridge University Press.

Beatty & Chaikin, (1990). The New Solar System 3rd Edition: Sky Publishing Corp.

Boss, A., (1998). Looking for Earths: John Wiley & Son's

Brady, J. E., (1990). General Chemistry - Principles and Structure: Wiley

Brahic, A. in Audouze/Israel, (1988). The Cambridge Atlas of Astronomy: Cambridge University Press

Brewer, D. (1992). Mercury and the Sun: Marshall Cavendish Corporation

Brewer, D. (1990). Saturn: Marshall Cavendish,

Brueton, D. (1991). Many Moons: Labyrinth Publishing S.A.

Burgess, E. & Dunne, J.A. (1978). The Voyage of Mariner 10: NASA California Institute of Technology

Capen, Dobbins, Parker, (1988). Observing and Photographing the Solar System: Willmann-Bell Publishing

Cazenave, in Audouze/Israel, (1988). The Cambridge Atlas of Astronomy: Cambridge University. Press

Chapman, Matthews, Vilas, (1988). Mercury: University of Arizona press.

Chemwiki: University of California: The Dynamical Chemistry E-textbook. Retrieved from chemwiki.ucdavis.edu.

Colin, Luhmann et al, (2002). The Solar Wind at 1 AU: EBSCO IND,

Comins, N.F. (1993). What if the Moon Didn't Exist? Harper Collins

Cooper H. S.F. JR (1993). The Evening Star: Harper Collins.

Cooper, Dobbins, Parker, (1988). Observing and Photographing the Solar System: Willmann-Bell

Croce, C. P. (2005). Mercury: NY Rosen Publishing Group

Daniels, P., (2005). The New Solar System, Ice Worlds, Moons, and Planets Redefined: National Geographic

Dunn, R., Wilkens, J. (2006). 300 Astronomical Objects: Firefly Books

Elkins-Tanton, L. T. (2006). The Sun, Mercury, and Venus: Chelsea House

Emiliani, C. (1992). Planet Earth: University of Cambridge

Emsley, (2000). The 13th Element, a Sorted Tale of Murder fire & Phosphorus: John Wiley & Sons

Enwright, M. (2014). The Sunday Edition: CBC Radio-2 episode 20:27 - March 7/2014

Fischer, D. E. (1987). The Birth of the Earth: Smyth-Sewn

Fitzpatrick, R; (n.d.) Newtonian Dynamics, Farside. ph.utexas.edu/teaching, Course notes University of Texas at Austin

French, B. M, Heiken G. H., Vaniman D. T. (1995). The Lunar Sourcebook: Cambridge University Press

Fritz, S. (2002). Understanding the New Solar System: Warner Books

Gilbert, H. & Gilbert-Smith, d. (1997). Gravity, the Glue of the Universe: Libraries Unlimited Inc.

Gonzalez & Richards (2004). The Privileged Planet: Regnery Publishing

Goss, T. (2003). Uranus, Neptune, and Pluto: Heinemann Library

Grinspoon, D. H. (1997). Venus Revealed: Addison

Grego, P. (2004). Moon Observer's Guide: Firefly Books

Harmon, D. (1999). Life out of Focus, Alzheimer's Disease and Related Disorders: Chelsea House

Harrington R. &Van Flandern, T. (1976). A Dynamical Investigation of the Conjecture that Mercury is a Escaped satellite of Venus: Icarus Vol. 28 Issue 4 August pgs. 435-440

Hartman, W. K. (1993). Moons and Planets: Wadsworth Inc.

Hayflick, L. (1994). How and Why We Age: Ballantine Books

Heiserman, D. L. (1991). Vital elements, Exploring Chemical Elements and Their Compounds:

Helmenstine, A. M. (2014). The Bohr Model of the Electron; retrieved from www.chemistry.about.com

Henbest, (1992). The Planets - Portraits of New Worlds; Viking

Hill, S. & Playfair, G. L. (1978). The Cycles of Heaven: Souvenir Press

Hunt, G. E. & Moore, P. (1994). Atlas of Neptune: Cambridge University Press

Israel, (1988). The Cambridge Atlas of Astronomy: Cambridge University. Press

Kern, M. (2005). Wisdom in the Body: North Atlantic Books

Kerrod, R. (2002). The Sun and Moon:

Kippenhahn, R. (1994). Discovering the |Secrets of the Sun: John Wiley & Sons

Kopal, Z. (1979). The Realm of the terrestrial Planets, The Bristol Institute of Physics

Lang, K. R. (2003). The Cambridge guide to the Solar System: Cambridge University Press

Lieber, A. L. Dr. (1996). How the Moon Affects You: Hastings House

Lunine, J. (1999). Earth, Evolution of a habitable World: Cambridge University Press

Lynch, J. (1999). The Moon, A&E/BBC Network

Mackenzie, D. (2003). Big Splat, the: John Wiley and Sons

MacCauley, J. F., (1977). Caloris & Orientale: Physics of the Earth and Planetary Interiors, 15 220-250

Macmillian (2004). Chemistry Foundations & Applications

Maor, E. (2000). Venus in Transit: Princeton University Press

Mason, (1988) Cambridge Atlas of Astronomy- Cambridge University Press

Mercury Magazine (2002). Vol.31.1

Miller, R. (2002). The Sun: Twenty- first Century Books.

Moore, P. (2001). On the Moon: Cassell and Co.

Moore, P. (2007). Moore on Mercury: Springer-Verlag

Morrison, D. (1993). Exploring Planetary Worlds: Scientific American Library

National Geographic Traveler's guide to the Planets:

Opperman, L.A. (2000). December, Cranial sutures as intra-membranous bone growth site: Developmental Dynamics Volume 29 issue 4, pgs 472-485.

Panno, J. (2011). Aging, Modern Theories and Therapies - Facts on File

Rhoades, R. and Pflanzer, R. (1996). Human Physiology 3rd Edition: Saunders College Publishing

Sharma, N. (2008). Parkinson's disease: Greenwood Press

Sheehan, W. (2004). The Transits of Venus, Prometheus Books

Sparrow, G. (2010). Destination Uranus, Neptune, and Pluto: The Rosen Publishing Group

Sprague A. L., Strom, R. G. (2003). Mercury, the Iron Planet: Springer C Praxis Publishing LTD

Strom, R. G. (1987). Mercury, the Elusive Planet: Smithsonian Institution Press

Stwertka, A. (1996). A Guide to the Elements 2nd Edition: Oxford University Press Guide to the Planets, the: Cambridge University Press

Thomas, (1988). The Cambridge Atlas of Astronomy, Cambridge University Press

Timelife, (1990). The Near Planets: Alexandria: Time-life books

Upledger, J. E. (1997). Your Inner Physician and You: North Atlantic Books

U.S. Health Department Website (2011). www.hhs.gov

Walsh-Shepard, D. (1994). Uranus: Franklin Watts.

Wagner, J. K. (1991). Introduction to the Solar System: Saunders College Pub., Whipple, F. L. (1981). Orbiting the Sun- Planets & Satellites of the Solar System, Harvard University Press.

About the Author

Toby L. Murray was born in Ann Arbor, Michigan and raised in Toronto, Canada. In 1994 he graduated from Sutherland-Chan Massage School and Teaching Clinic. He presently practices as a registered massage therapist in Toronto while working toward a degree in Astronomy at York University.

To Contact Toby L. Murray RMT
- <u>tlmrmt@hotmail.com</u>
- (Twitter) #Mr_Menopause
- menopauseandmercury@wordblog.com

www.ingramcontent.com/pod-product-compliance
Lightning Source LLC
Chambersburg PA
CBHW021026180526
45163CB00005B/2138